Intermediate 1 | Units 1, 2 & 3
Mathematics

2004 Exam
Paper 1 (Non-calculator)
Paper 2

2005 Exam
Paper 1 (Non-calculator)
Paper 2

2006 Exam
Paper 1 (Non-calculator)
Paper 2

2007 Exam
Paper 1 (Non-calculator)
Paper 2

2008 Exam
Paper 1 (Non-calculator)
Paper 2

Leckie×Leckie

First exam published in 2004.
Published by Leckie & Leckie Ltd, 3rd Floor, 4 Queen Street, Edinburgh EH2 1JE
tel: 0131 220 6831 fax: 0131 225 9987 enquiries@leckieandleckie.co.uk www.leckieandleckie.co.uk

ISBN 978-1-84372-651-7

A CIP Catalogue record for this book is available from the British Library.

Leckie & Leckie is a division of Huveaux plc.

Leckie & Leckie is grateful to the copyright holders, as credited at the back of the book, for permission to use their material.
Every effort has been made to trace the copyright holders and to obtain their permission for the use of copyright material.
Leckie & Leckie will gladly receive information enabling them to rectify any error or omission in subsequent editions.

[BLANK PAGE]

FOR OFFICIAL USE

Total mark

X100/101

NATIONAL
QUALIFICATIONS
2004

FRIDAY, 21 MAY
1.00 PM – 1.35 PM

MATHEMATICS
INTERMEDIATE 1
Units 1, 2 and 3
Paper 1
(Non-calculator)

Fill in these boxes and read what is printed below.

Full name of centre

Town

Forename(s)

Surname

Date of birth
Day Month Year

Scottish candidate number

Number of seat

1   You may **NOT** use a calculator.

2   Write your working and answers in the spaces provided. Additional space is provided at the end of this question-answer book for use if required. If you use this space, write clearly the number of the question involved.

3   Full credit will be given only where the solution contains appropriate working.

4   Before leaving the examination room you must give this book to the invigilator. If you do not you may lose all the marks for this paper.

SCOTTISH
QUALIFICATIONS
AUTHORITY

FORMULAE LIST

Circumference of a circle: $C = \pi d$
Area of a circle: $A = \pi r^2$

Theorem of Pythagoras:

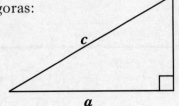

$a^2 + b^2 = c^2$

Trigonometric ratios
in a right angled
triangle:

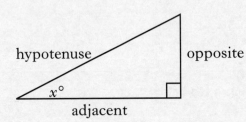

$$\tan x° = \frac{\text{opposite}}{\text{adjacent}}$$

$$\sin x° = \frac{\text{opposite}}{\text{hypotenuse}}$$

$$\cos x° = \frac{\text{adjacent}}{\text{hypotenuse}}$$

ALL questions should be attempted.

Marks

1. Work out the answers to the following.

   (a) 30% of £230

   1

   (b) $\frac{4}{7}$ of 105

   1

   (c) 380 − 20 × 9

   1

2. A cooker can be bought by paying a deposit of £59 followed by 12 instalments of £45.

   Calculate the total price of the cooker.

   2

[Turn over

Marks

3. Calculate the volume of this cuboid.

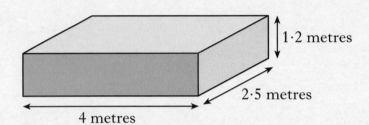

1·2 metres

2·5 metres

4 metres

2

4. The temperatures, in degrees Celsius, at noon for the first ten days in January at Invergow were:

$$-3 \quad 0 \quad -2 \quad 2 \quad -1 \quad -4 \quad -5 \quad -3 \quad 1 \quad 3.$$

Calculate

(*a*) the median temperature;

2

(*b*) the range.

2

Marks

4. (continued)

(*c*) The corresponding values of the median and the range for Abergrange are 2 °C and 5 °C respectively.

Make **two** comments comparing the temperatures in Invergow and Abergrange.

2

5. Solve algebraically the equation

$$11 + 5x = 2x + 29.$$

3

[Turn over

Marks

6. A shop sells artificial flowers.
The prices of individual flowers are shown below.

Variety	Price
Carnation	£2
Daffodil	£3·50
Lily	£4
Iris	£3
Rose	£4·50

Zara wants to
- buy 3 flowers
- choose 3 different varieties
- spend a **minimum** of £10.

One combination of flowers that Zara can buy is shown in the table below.

Carnation	Daffodil	Lily	Iris	Rose	Total Price
		✓	✓	✓	£11·50

Complete the table to show **all** the possible combinations that Zara can buy.   3

Marks

7. An Internet provider has a customer helpline.
The length of each telephone call to the helpline was recorded one day.
The results are shown in the frequency table below.

Length of call (to nearest minute)	Frequency	Length of call × Frequency
1	15	15
2	40	80
3	26	78
4	29	116
5	49	
6	41	
	Total = 200	Total =

(a) Complete the table above and find the mean length of call.

(b) Write down the modal length of call.

3

1

[Turn over

Marks

8. (*a*) Complete the table below for $y = 3 - x$.

x	-2	2	7
y			

2

(*b*) Draw the line $y = 3 - x$ on the grid.

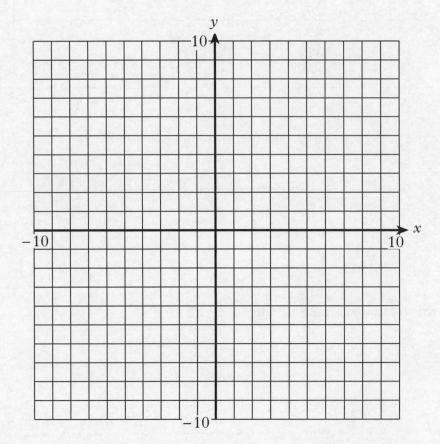

2

Marks

9. One billion is 1000 million.

A country borrows 2×10^{10} dollars.

How many billions of dollars is this?

3

10. Evaluate $\dfrac{2xy}{z}$ when $x = -5$, $y = 6$ and $z = -4$.

[END OF QUESTION PAPER]

3

DO NOT
WRITE IN
THIS
MARGIN

ADDITIONAL SPACE FOR ANSWERS

FOR OFFICIAL USE

Total mark

X100/103

NATIONAL
QUALIFICATIONS
2004

FRIDAY, 21 MAY
1.55 PM – 2.50 PM

MATHEMATICS
INTERMEDIATE 1
Units 1, 2 and 3
Paper 2

Fill in these boxes and read what is printed below.

Full name of centre

Town

Forename(s)

Surname

Date of birth

Day Month Year   Scottish candidate number   Number of seat

1 **You may use a calculator.**

2 Write your working and answers in the spaces provided. Additional space is provided at the end of this question-answer book for use if required. If you use this space, write clearly the number of the question involved.

3 Full credit will be given only where the solution contains appropriate working.

4 Before leaving the examination room you must give this book to the invigilator. If you do not you may lose all the marks for this paper.

SCOTTISH
QUALIFICATIONS
AUTHORITY

FORMULAE LIST

Circumference of a circle: $C = \pi d$

Area of a circle: $A = \pi r^2$

Theorem of Pythagoras:

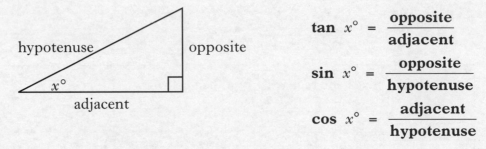

$a^2 + b^2 = c^2$

Trigonometric ratios
in a right angled
triangle:

$$\tan x° = \frac{\text{opposite}}{\text{adjacent}}$$

$$\sin x° = \frac{\text{opposite}}{\text{hypotenuse}}$$

$$\cos x° = \frac{\text{adjacent}}{\text{hypotenuse}}$$

Marks

ALL questions should be attempted.

1. 2000 tickets are sold for a raffle in which the star prize is a television.
   Kirsty buys 10 tickets for the raffle.
   What is the probability that she wins the star prize?

   1

2. (*a*) On the grid below, plot the points A(–3, 4), B(2, 4) and C(6, –5).

   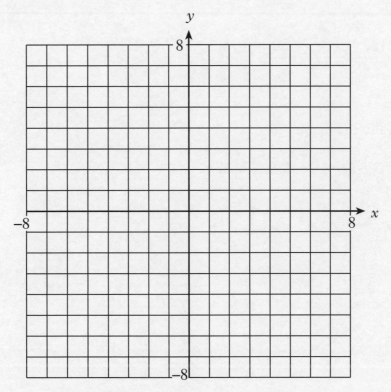

   2

   (*b*) Plot the point D so that shape ABCD is a kite.
   Write down the coordinates of point D.

   2

Page three

[Turn over

Marks

3. An overnight train left London at 2040 and reached Inverness at 0810 the next day.

   The distance travelled by the train was 552 miles.

   Calculate the average speed of the train.

   3

4. Solve algebraically the inequality

   $$8n - 3 < 37.$$

   2

Page four

Marks

5. The scattergraph shows the age and mileage of cars in a garage.

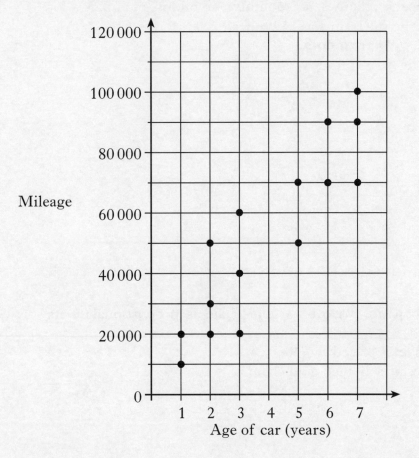

(*a*) Draw a line of best fit through the points on the graph.   **1**

(*b*) Use your line of best fit to estimate the mileage of a 4 year old car.

1

6. (*a*) Multiply out the brackets and simplify

$$2(4 - t) + 5t.$$

2

(*b*) Factorise $10y - 35.$

2

Marks

7. Ryan wants to take out a life insurance policy.

The insurance company charges a monthly premium of £2·50 for each £1000 of cover.

Ryan can afford to pay £90 per month.

How much cover can he get?

2

8. (*a*) In a jewellery shop the price of a gold chain is proportional to its length.

A 16 inch gold chain is priced at £40.

Calculate the price of a 24 inch gold chain.

2

(*b*) The gold chains are displayed diagonally on a **square** board of side 20 inches.

The longest chain stretches from corner to corner.

Calculate the length of the longest chain.

Do not use a scale drawing.

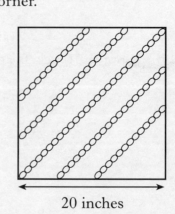

20 inches

3

Marks

9. Andy buys a bottle of aftershave in Spain for 38·50 euros.

The same bottle of aftershave costs £25·99 in Scotland.

The exchange rate is £1 = 1·52 euros.

Does he save money by buying the aftershave in Spain?

Explain your answer.

3

10. The front of the tent shown below is an isosceles triangle.

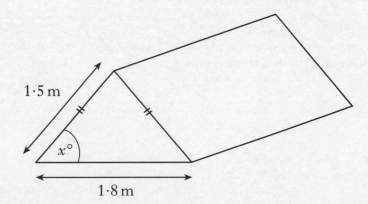

The size of the angle between the side and the bottom of the tent is $x°$.

Calculate x.

4

Page seven

[Turn over

Marks

11. The graph below shows how the rate of interest for a savings account with the Clydeside Bank changed during 2002.

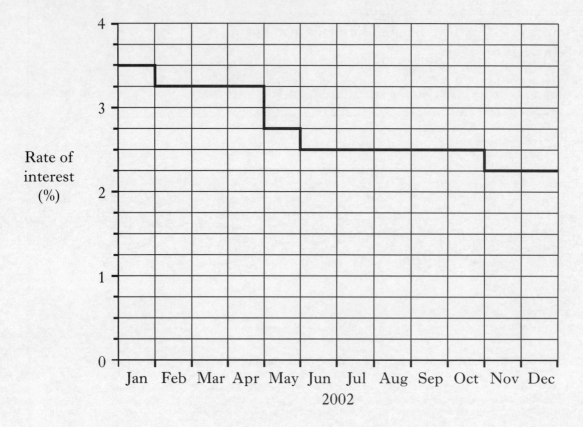

Rate of interest (%)

Jan Feb Mar Apr May Jun Jul Aug Sep Oct Nov Dec
2002

(*a*) What was the interest rate in March?

1

(*b*) Describe the trend of the interest rate during the year.

1

(*c*) In June £1400 was invested in this account.
How much interest was due after 3 months?

4

Marks

12. The minimum velocity v metres per second, allowed at the top of a loop in a roller coaster, is given by the formula

$$v = \sqrt{gr}$$

where r metres is the radius of the loop.

Calculate the value of v when $g = 9\cdot81$ and $r = 9$.

3

13. 40 people were asked whether they preferred tea or coffee.

18 of them said they preferred coffee.

What percentage said they preferred coffee?

3

[Turn over

Marks

14. The diagram below shows a rectangular door with a window.

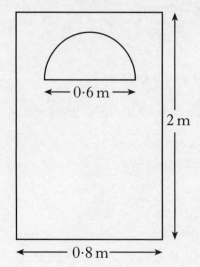

The window is in the shape of a semi-circle and is made of glass.

The rest of the door is made of wood.

Calculate the area of the wooden part of the door.

Give your answer in square metres correct to two decimal places.

5

[END OF QUESTION PAPER]

[BLANK PAGE]

FOR OFFICIAL USE

Total mark

X100/101

NATIONAL QUALIFICATIONS 2005

FRIDAY, 20 MAY 1.00 PM – 1.35 PM

MATHEMATICS INTERMEDIATE 1
Units 1, 2 and 3
Paper 1
(Non-calculator)

Fill in these boxes and read what is printed below.

Full name of centre

Town

Forename(s)

Surname

Date of birth
Day Month Year

Scottish candidate number

Number of seat

1 **You may NOT use a calculator.**

2 Write your working and answers in the spaces provided. Additional space is provided at the end of this question-answer book for use if required. If you use this space, write clearly the number of the question involved.

3 Full credit will be given only where the solution contains appropriate working.

4 Before leaving the examination room you must give this book to the invigilator. If you do not you may lose all the marks for this paper.

SCOTTISH
QUALIFICATIONS
AUTHORITY

FORMULAE LIST

Circumference of a circle: $C = \pi d$

Area of a circle: $A = \pi r^2$

Theorem of Pythagoras:

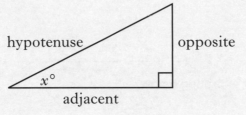

$$a^2 + b^2 = c^2$$

Trigonometric ratios
in a right angled
triangle:

$$\tan x^\circ = \frac{\text{opposite}}{\text{adjacent}}$$

$$\sin x^\circ = \frac{\text{opposite}}{\text{hypotenuse}}$$

$$\cos x^\circ = \frac{\text{adjacent}}{\text{hypotenuse}}$$

Marks

ALL questions should be attempted.

1. (*a*) Find $6{\cdot}17 - 2{\cdot}3$.

1

(*b*) Find 75% of £1200.

1

2. Joyce is going on holiday. She must be at the airport by 1.20 pm. It takes her 4 hours 30 minutes to travel from home to the airport. What is the latest time that she should leave home for the airport?

1

[Turn over

Marks

3. A regular polygon is a shape with three or more equal sides.

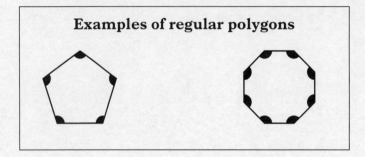

A rule used to calculate the size, in degrees, of each angle in a regular polygon is:

Size of each angle = 180 − (360 ÷ number of sides)

Calculate the size of each angle in the regular polygon below.

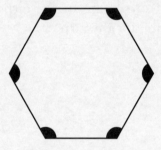

Do not measure with a protractor.

You must show your working.

2

Marks

4. The number of peas counted in each of 100 pea pods
is shown in this frequency table.

Peas in pod	Frequency	Peas in pod × Frequency
3	5	15
4	10	40
5	28	140
6	36	216
7	12	
8	9	
	Total = 100	Total =

Complete the table above **and** calculate the mean number of peas in a pod.

3

5. Solve algebraically the equation

$$11a - 8 = 37 + 6a.$$

3

[Turn over

Marks

6. Anwar wants to buy some accessories for his computer.
He sees this advert for Cathy's Computers.

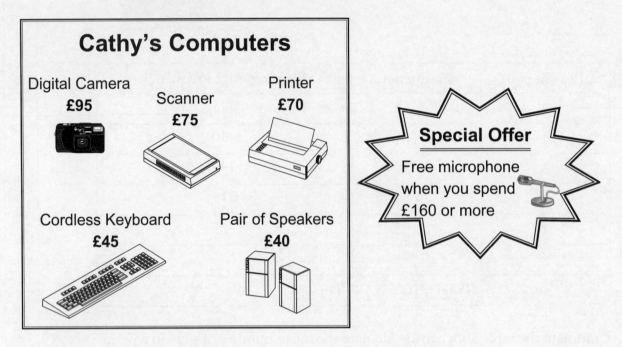

Anwar wants to spend enough to get the free microphone.

He can afford to spend a maximum of £200.

He does not want to buy more than one of each accessory.

One combination of accessories that Anwar can buy is shown in the table below.

Digital Camera £95	Scanner £75	Printer £70	Cordless Keyboard £45	Pair of Speakers £40	Total Value
	✓	✓		✓	£185

Complete the table to show **all** possible combinations that Anwar can buy.

3

Marks

7. (*a*) Complete the table below for $y = -2x + 5$.

x	-2	0	4
y			

2

(*b*) Draw the line $y = -2x + 5$ on the grid.

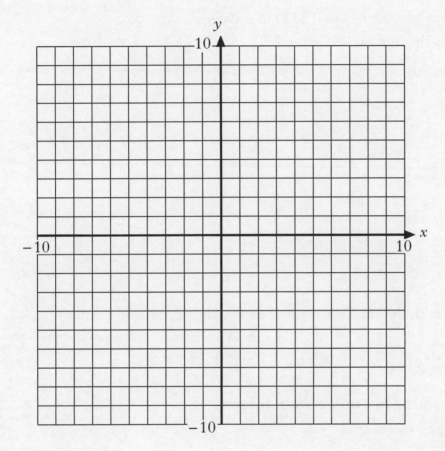

2

[Turn over

Marks

8. (a) While in New York, Martin changed £50 into US dollars.
   The exchange rate was £1 = $1·62.
   How many US dollars did Martin receive for £50?

2

(b) A few days later he received $320 in exchange for £200.
   What was the new exchange rate?

2

9. (a) Write $\dfrac{7}{1000}$ as a decimal.

1

(b) Starting with the smallest, write the following numbers in order.

$$\frac{7}{1000}, \qquad 0\cdot069, \qquad 7\cdot1 \times 10^{-4}$$

Show working to explain your answer.

3

Marks

10. In a **magic square**, the numbers in each row, each column and each diagonal add up to the same **magic total**.

In this magic square the **magic total** is 3.

−2	5	0
3	1	−1
2	−3	4

(a)

−4	3	−2
1	−1	−3
0	−5	2

This is another magic square.
What is its **magic total**?

1

(b) Complete this **magic square**.

1		
	−2	
−3		−5

3

[END OF QUESTION PAPER]

ADDITIONAL SPACE FOR ANSWERS

ADDITIONAL SPACE FOR ANSWERS

ADDITIONAL SPACE FOR ANSWERS

FOR OFFICIAL USE

Total mark

X100/103

NATIONAL
QUALIFICATIONS
2005

FRIDAY, 20 MAY
1.55 PM – 2.50 PM

MATHEMATICS
INTERMEDIATE 1
Units 1, 2 and 3
Paper 2

Fill in these boxes and read what is printed below.

Full name of centre

Town

Forename(s)

Surname

Date of birth
Day Month Year   Scottish candidate number   Number of seat

1   **You may use a calculator.**

2   Write your working and answers in the spaces provided. Additional space is provided at the end of this question-answer book for use if required. If you use this space, write clearly the number of the question involved.

3   Full credit will be given only where the solution contains appropriate working.

4   Before leaving the examination room you must give this book to the invigilator. If you do not you may lose all the marks for this paper.

SCOTTISH
QUALIFICATIONS
AUTHORITY

FORMULAE LIST

Circumference of a circle: $C = \pi d$
Area of a circle: $A = \pi r^2$

Theorem of Pythagoras:

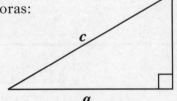

$$a^2 + b^2 = c^2$$

Trigonometric ratios
in a right angled
triangle:

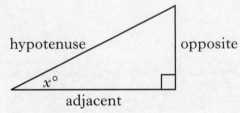

$$\tan x^\circ = \frac{\text{opposite}}{\text{adjacent}}$$

$$\sin x^\circ = \frac{\text{opposite}}{\text{hypotenuse}}$$

$$\cos x^\circ = \frac{\text{adjacent}}{\text{hypotenuse}}$$

Marks

ALL questions should be attempted.

1. Calculate the volume of the cube below.

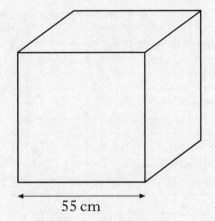

55 cm

Round your answer to the nearest thousand cubic centimetres.

2

2. Claire sells cars.

She is paid £250 per month plus 3% commission on her sales.

How much is she paid in a month when her sales are worth £72 000?

2

[Turn over

Marks

3. A group of students visit a theme park.

The graph below shows their journey.

They set off from the college at 9 am and arrive back at 4 pm.

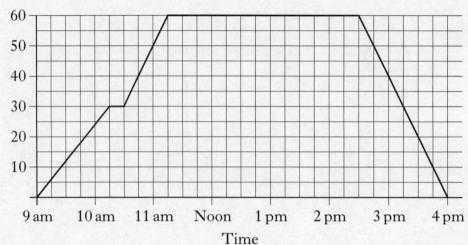

(*a*) How long did the students spend at the theme park?

1

(*b*) Calculate the average speed, in miles per hour, of the students' return journey.

3

4. Solve algebraically the inequality

$$3t + 4 > 28.$$

2

Marks

5. The stem and leaf diagram below shows the ages of the players in the Kestrels rugby team.

AGES
Kestrels

```
1 | 9
2 | 1 3 4 7 9
3 | 0 2 4 5 5 5 8 9
4 | 1
```

2 | 1 represents 21 years

(*a*) What age is the oldest player?

1

(*b*) Calculate the range of ages.

2

The stem and leaf diagram below shows the ages of both the Kestrels and the Falcons rugby teams.

AGES

Falcons   **Kestrels**

```
        9 9| 1 |9
8 7 7 6 3 2 1 1 0| 2 |1 3 4 7 9
        8 6 4 3| 3 |0 2 4 5 5 5 8 9
                | 4 |1
```

2 | 1 represents 21 years

(*c*) Compare the ages of the two teams. Comment on any difference.

1

[Turn over

[X100/103]   *Page five*

Marks

6. (*a*) Multiply out the brackets and simplify

$$11n + 4(7 - 2n).$$

2

(*b*) Factorise $\qquad 15 + 6x.$

2

7. The scores of 12 golfers in a competition were as follows.

67	70	68	75	71	70
70	75	76	75	74	75

(*a*) Find the modal score.

1

(*b*) Find the median score.

2

(*c*) Find the probability of choosing a golfer from this group with a score of 70.

1

Marks

8. 60 workers in a factory voted on a new pay deal.

    42 of them voted to accept the deal.

    What percentage voted to accept the deal?

3

9. The pie chart shows the different sizes of eggs laid by a flock of hens.

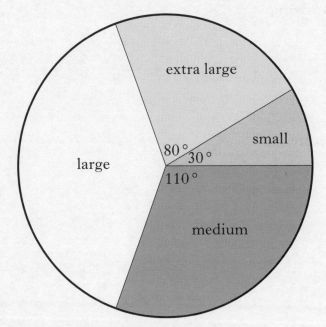

The flock of hens laid 1260 eggs.

How many of the eggs were large?

3

[Turn over

Marks

10. A rectangular shelf is supported by brackets as shown.
Each bracket is a right angled triangle.

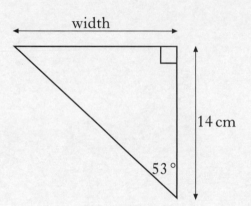

width

14 cm

53°

Calculate the width of this bracket.

Give your answer correct to one decimal place.

Do not use a scale drawing.

4

Marks

11. The diagram below shows a speedway track.

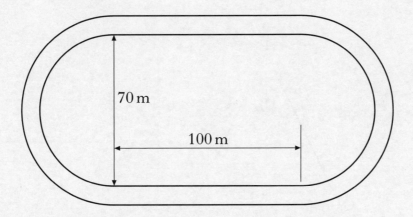

70 m

100 m

The straights are each 100 metres long.

The bends are semi-circles as shown.

Calculate the perimeter of the inside of the track.

4

12. Use the formula below to find the value of A when $b = 2{\cdot}4$ and $c = 5$.

$$A = 3bc^2$$

3

[Turn over

Marks

13. PQRS is a rhombus.

The diagonals PR and QS are 15 centimetres and 8 centimetres long as shown.

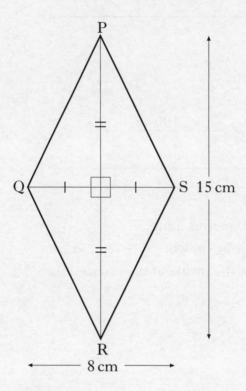

Calculate the length of side PQ.

Do not use a scale drawing.

3

14. Margaret is recovering from an operation.

She needs to take 4 tablets each day for a year.

The tablets are supplied in boxes of 200.

Each box costs £6·50.

How much does it cost for the year's supply?

3

Marks

15. The diagram below shows a plan of a patio.

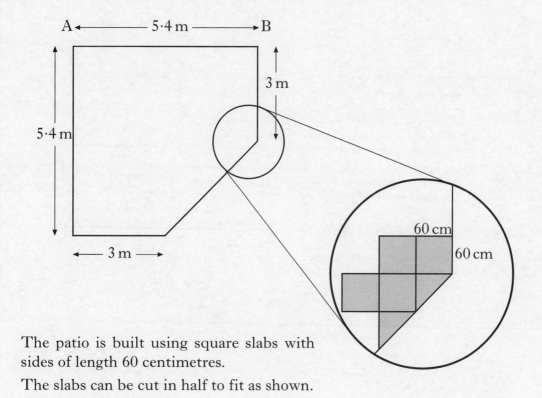

The patio is built using square slabs with sides of length 60 centimetres.

The slabs can be cut in half to fit as shown.

(*a*) How many slabs fit exactly along edge AB?

1

(*b*) How many slabs are needed altogether to build the patio?

4

[END OF QUESTION PAPER]

ADDITIONAL SPACE FOR ANSWERS

[BLANK PAGE]

FOR OFFICIAL USE

Total mark

X100/101

NATIONAL
QUALIFICATIONS
2006

FRIDAY, 19 MAY
1.00 PM – 1.35 PM

MATHEMATICS
INTERMEDIATE 1
Units 1, 2 and 3
Paper 1
(Non-calculator)

Fill in these boxes and read what is printed below.

Full name of centre

Town

Forename(s)

Surname

Date of birth
Day Month Year

Scottish candidate number

Number of seat

1 **You may NOT use a calculator.**

2 Write your working and answers in the spaces provided. Additional space is provided at the end of this question-answer book for use if required. If you use this space, write clearly the number of the question involved.

3 Full credit will be given only where the solution contains appropriate working.

4 Before leaving the examination room you must give this book to the invigilator. If you do not you may lose all the marks for this paper.

SCOTTISH
QUALIFICATIONS
AUTHORITY

FORMULAE LIST

Circumference of a circle: $C = \pi d$

Area of a circle: $A = \pi r^2$

Theorem of Pythagoras:

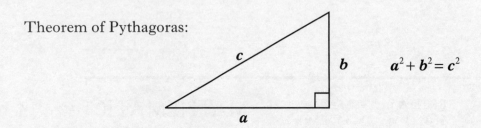

$$a^2 + b^2 = c^2$$

Trigonometric ratios
in a right angled
triangle:

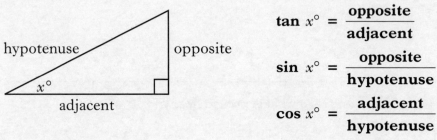

$$\tan x^\circ = \frac{\text{opposite}}{\text{adjacent}}$$

$$\sin x^\circ = \frac{\text{opposite}}{\text{hypotenuse}}$$

$$\cos x^\circ = \frac{\text{adjacent}}{\text{hypotenuse}}$$

Marks

ALL questions should be attempted.

1. Find $5 \cdot 42 - 1 \cdot 8$.

1

2. A tree surgeon uses this rule to work out his charge in pounds for uprooting and removing trees.

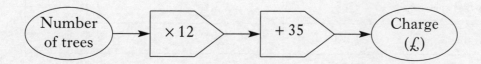

How much would he charge to uproot and remove 11 trees?

2

[Turn over

Marks

3. Paula runs a 1500 metre race at an average speed of 6 metres per second.

   How long does she take to run the race?

   Give her time in minutes and seconds.

3

4. The table below shows insurance premiums for holidays abroad.

	INSURANCE PREMIUM per adult		
	Europe	Worldwide	Winter Sports
Up to 8 days	£15	£30	£40
9–17 days	£20	£40	£55
18–26 days	£30	£60	£80

Child premium (0–15 years) is 70% of the adult premium.

Mr and Mrs Fleming and their 5 year old son go to the USA for a three week holiday in July.

Find the **total** insurance premium for the family.

3

Marks

5. The hire purchase price of this camcorder is £499.

£499

£85 deposit
followed by 9 equal payments

How much will each payment be?

3

6. Solve algebraically the equation

$$5n + 9 = 51 - 2n.$$

3

[Turn over

Marks

7. (*a*) Complete the table below for $y = 2 + 3x$.

x	-3	0	2
y			

2

(*b*) Draw the line $y = 2 + 3x$ on the grid.

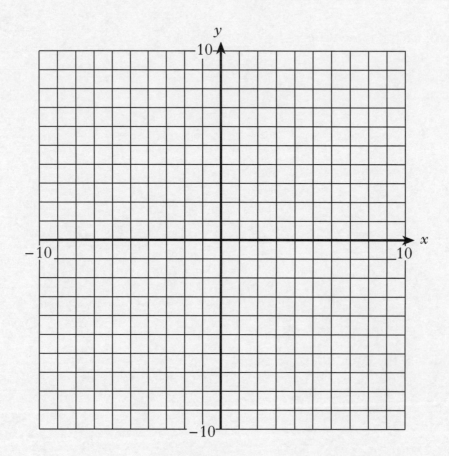

2

Marks

8. A television programme has a phone-in to raise money for charity.

The calls cost 70 pence per minute.

The charity receives $\frac{3}{5}$ of the cost of each call.

How much money will the charity receive from a call which lasts $2\frac{1}{2}$ minutes?

3

9. Use the formula below to find the value of I when $P = 144$ and $R = 4$.

$$I = \sqrt{\frac{P}{R}}$$

3

[Turn over for Question 10 on *Page eight*

Marks

10. This is a number cell.

	1st	2nd	3rd	4th
	3	−2	1	−1

- 1st number + 2nd number = 3rd number $\quad$ 3 + (−2) = 1
- 2nd number + 3rd number = 4th number $\quad\quad$ (−2) + 1 = −1

(*a*) Complete this number cell.

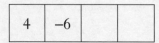

1

(*b*) Complete this number cell.

2

(*c*) Complete this number cell.

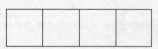

2

YOU MAY USE THE BLANK NUMBER CELLS BELOW FOR WORKING IF YOU WISH.

[END OF QUESTION PAPER]

ADDITIONAL SPACE FOR ANSWERS

DO NOT
WRITE IN
THIS
MARGIN

ADDITIONAL SPACE FOR ANSWERS

Page ten

ADDITIONAL SPACE FOR ANSWERS

[BLANK PAGE]

FOR OFFICIAL USE

Total mark

X100/103

NATIONAL
QUALIFICATIONS
2006

FRIDAY, 19 MAY
1.55 PM – 2.50 PM

MATHEMATICS
INTERMEDIATE 1
Units 1, 2 and 3
Paper 2

Fill in these boxes and read what is printed below.

Full name of centre

Town

Forename(s)

Surname

Date of birth

Day Month Year   Scottish candidate number   Number of seat

1 **You may use a calculator.**

2 Write your working and answers in the spaces provided. Additional space is provided at the end of this question-answer book for use if required. If you use this space, write clearly the number of the question involved.

3 Full credit will be given only where the solution contains appropriate working.

4 Before leaving the examination room you must give this book to the invigilator. If you do not you may lose all the marks for this paper.

SCOTTISH
QUALIFICATIONS
AUTHORITY

FORMULAE LIST

Circumference of a circle: $C = \pi d$

Area of a circle: $A = \pi r^2$

Theorem of Pythagoras:

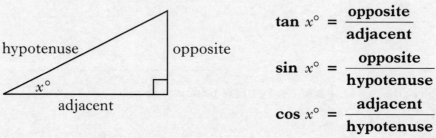

$a^2 + b^2 = c^2$

Trigonometric ratios in a right angled triangle:

$$\tan x° = \frac{\text{opposite}}{\text{adjacent}}$$

$$\sin x° = \frac{\text{opposite}}{\text{hypotenuse}}$$

$$\cos x° = \frac{\text{adjacent}}{\text{hypotenuse}}$$

Marks

ALL questions should be attempted.

1. During a holiday in Mexico, Lee changed £650 into pesos.

   The exchange rate was £1 = 19·13 pesos.

   How many pesos did Lee receive for £650?

   Round off your answer to the nearest ten pesos.

   2

2. Light travels one mile in about 0·000 005 4 seconds.

   Write this time in standard form.

   2

   [Turn over

Marks

3. Solve algebraically the inequality

$$4t - 7 > 29.$$

2

4. The number of bricks needed to build a wall is proportional to the area of the wall.

A wall with an area of 4 square metres needs 260 bricks.

How many bricks are needed for a wall with an area of 7 square metres?

2

Marks

5. A group of 40 students sit a test.

   The marks scored by the students in the test are shown in the frequency table below.

Mark	Frequency
14	6
15	10
16	7
17	7
18	5
19	3
20	2

   (a) Write down the modal mark.

   1

   (b) Find the probability of choosing a student from this group with a mark of 19.

   1

   (c) Complete the table below and calculate the mean mark for the group.

Mark	Frequency	Mark x Frequency
14	6	84
15	10	150
16	7	112
17	7	119
18	5	
19	3	
20	2	
	Total = 40	Total =

   3

 [Turn over

Marks

6. A water tank is 50 centimetres wide, 1·2 metres long and 40 centimetres high. Calculate its volume.

Give your answer in litres.
(1 litre = 1000 cubic centimetres.)

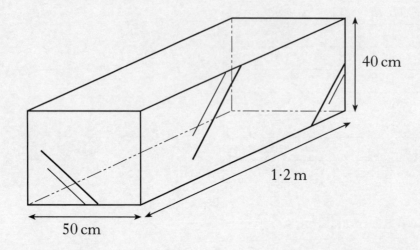

40 cm

1·2 m

50 cm

3

7. (*a*) Multiply out the brackets and simplify

$$3y + 2(x - 4y).$$

2

(*b*) Factorise $\quad\quad 8d + 12.$

2

Marks

8. Every morning for one week, Wellburgh Council carried out a traffic survey at a busy junction.

The number of cars waiting to turn right at the junction was counted every five minutes between 8 am and 9 am.

On Monday morning the results were:

    10   14   17   12   14   11   13   7   8   7   6   2.

Calculate:

(a) the median;

2

(b) the range.

2

On Saturday morning, the median was 6 and the range was 8.

(c) Make **two** comments comparing the number of cars waiting to turn right at the junction on Monday morning and Saturday morning.

2

[Turn over

Marks

9. Stephen is playing snooker.

    The diagram below shows the positions of three balls on the table.

    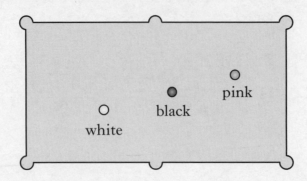

    Stephen plays the white ball, W.

    It bounces off the side of the table at X and hits the pink ball, P.

    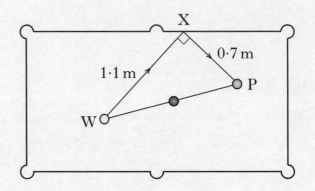

    • Distance WX is 1·1 metres
    • Distance XP is 0·7 metres
    • Angle WXP is 90°

    Calculate distance WP.

    Do not use a scale drawing.

    3

Marks

10. The table below shows the stopping distances of a car, when the brakes are applied, at different speeds.

Speed (miles per hour)	0	10	20	30	40
Stopping distance (feet)	0	15	40	75	120

On the grid below, draw a **line** graph to show this information.

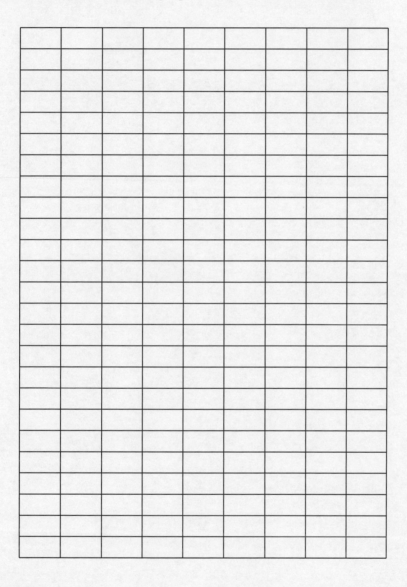

4

[Turn over

DO NOT WRITE IN THIS MARGIN

Marks

11. Ralph invests £2600 in a building society account.

The rate of interest is 4·5% per annum.

Calculate the interest he should receive after 8 months.

3

Marks

12. A road bridge can be raised in the **centre** to allow ships to pass through.

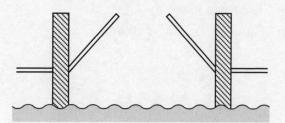

The moveable sections of the bridge are:

- 10 metres above the water level
- 40 metres long altogether.

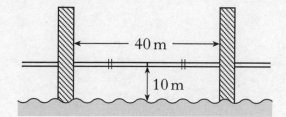

The moveable sections are raised through 50° to allow a ship to pass through.

Calculate the height of the point P above the water level.

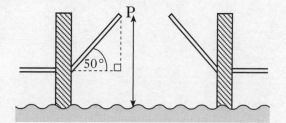

Do not use a scale drawing.

5

DO NOT WRITE IN THIS MARGIN

Marks

13. Andrew designs a website to advertise his hotel.

In the first month he has 250 visitors to his site.

The following month he has 300 visitors.

Calculate the percentage increase in the number of visitors. **4**

Marks

14. The diagram below shows the wall at the start of a tunnel.

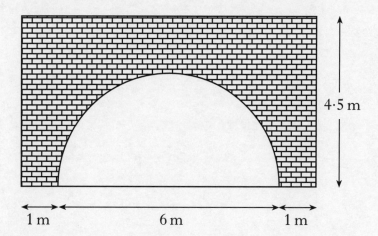

The wall is in the shape of a rectangle with a semi-circular space for the tunnel.

Calculate the area of the wall in square metres.

Give your answer correct to one decimal place.

5

[*END OF QUESTION PAPER*]

DO NOT
WRITE IN
THIS
MARGIN

ADDITIONAL SPACE FOR ANSWERS

[BLANK PAGE]

FOR OFFICIAL USE

Total
mark

X100/101

NATIONAL
QUALIFICATIONS
2007

TUESDAY, 15 MAY
1.00 PM – 1.35 PM

MATHEMATICS
INTERMEDIATE 1
Units 1, 2 and 3
Paper 1
(Non-calculator)

Fill in these boxes and read what is printed below.

Full name of centre

Town

Forename(s)

Surname

Date of birth

Day Month Year

Scottish candidate number

Number of seat

1 **You may NOT use a calculator.**

2 Write your working and answers in the spaces provided. Additional space is provided at the end of this question-answer book for use if required. If you use this space, write clearly the number of the question involved.

3 Full credit will be given only where the solution contains appropriate working.

4 Before leaving the examination room you must give this book to the invigilator. If you do not you may lose all the marks for this paper.

FORMULAE LIST

Circumference of a circle: $C = \pi d$
Area of a circle: $A = \pi r^2$

Theorem of Pythagoras:

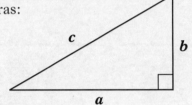

$a^2 + b^2 = c^2$

Trigonometric ratios
in a right angled
triangle:

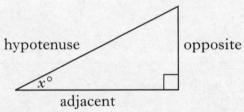

$$\tan x° = \frac{\text{opposite}}{\text{adjacent}}$$

$$\sin x° = \frac{\text{opposite}}{\text{hypotenuse}}$$

$$\cos x° = \frac{\text{adjacent}}{\text{hypotenuse}}$$

Marks

ALL questions should be attempted.

1. (a) Find 8·52 + 10·7.

1

(b) Find 3·76 ÷ 8.

1

(c) Change 0·057 into a fraction.

1

(d) Find 90% of £320.

2

2. Shona wants to insure her jewellery for £8000.

The insurance company charges an annual premium of £7·65 for each £1000 insured.

Work out Shona's annual premium.

2

Marks

3. Solve algebraically the inequality

$$7a + 6 < 69.$$

2

4. The number of minutes that patients had to sit in the waiting room before seeing their doctor was recorded one day.

The results are shown in the frequency table below.

Number of minutes	Frequency	Number of minutes × Frequency
5	4	20
6	7	42
7	8	56
8	13	104
9	12	
10	6	
	Total = 50	Total =

Complete the table above **and** find the mean number of minutes.

3

Marks

5. (*a*) Complete the table below for $y = 4x - 3$.

x	−1	0	1	3
y			1	

2

(*b*) Draw the line $y = 4x - 3$ on the grid.

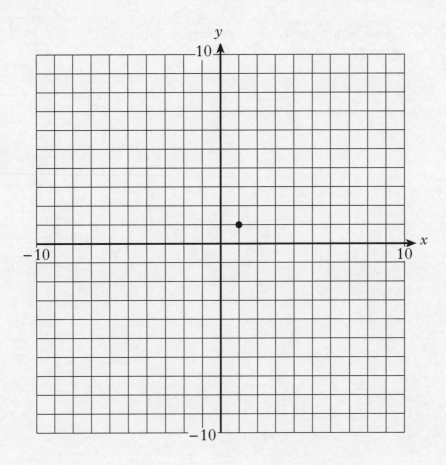

2

[Turn over

Marks

6. Shown below is a container in the shape of a cuboid.

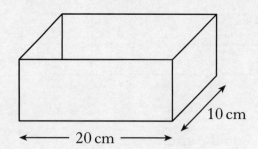

20 cm

10 cm

When full, the container holds 1600 cubic centimetres of water.

Work out the height of the container.

3

7. Work out the answers to the following.

(*a*) $2 \times (-2) \times 2$

1

(*b*) $11 - (-6)$

1

Marks

8. Naveed has six electrical appliances in his student lodgings.

The power, in watts, used by each appliance is shown below.

Lamp 100 watts

Computer 200 watts

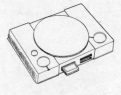

Games Machine
400 watts

Microwave 700 watts

Heater 1000 watts

Kettle 2300 watts

Naveed uses a 4-way extension lead for the appliances.

The instructions state that the maximum power used through the extension lead should not be more than 3000 watts.

One combination of **four** appliances that Naveed can safely use through the extension lead is shown in the table below.

Lamp 100 watts	Computer 200 watts	Games Machine 400 watts	Microwave 700 watts	Heater 1000 watts	Kettle 2300 watts	Total Watts
✓	✓	✓		✓		1700

Complete the table to show **all** the possible combinations of **four** appliances that Naveed can safely use through the extension lead.

3

[Turn over for Questions 9 and 10 on *Page eight*

9. The formula for the area of a trapezium is

$$A = \tfrac{1}{2}h(a + b)$$

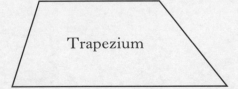

Trapezium

Find A when $a = 11$, $b = 7$ and $h = 6$.

3

10. Black and white counters are placed in two bags as shown below.

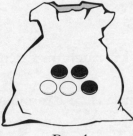

Bag 1

Bag 2

One counter is selected at random from **each** bag.

Which bag gives a greater probability of selecting a black counter?

Explain your answer.

3

[END OF QUESTION PAPER]

ADDITIONAL SPACE FOR ANSWERS

ADDITIONAL SPACE FOR ANSWERS

ADDITIONAL SPACE FOR ANSWERS

[BLANK PAGE]

FOR OFFICIAL USE

Total mark

X100/103

NATIONAL
QUALIFICATIONS
2007

TUESDAY, 15 MAY
1.55 PM – 2.50 PM

MATHEMATICS
INTERMEDIATE 1
Units 1, 2 and 3
Paper 2

Fill in these boxes and read what is printed below.

Full name of centre

Town

Forename(s)

Surname

Date of birth
Day Month Year

Scottish candidate number

Number of seat

1   **You may use a calculator.**

2   Write your working and answers in the spaces provided. Additional space is provided at the end of this question-answer book for use if required. If you use this space, write clearly the number of the question involved.

3   Full credit will be given only where the solution contains appropriate working.

4   Before leaving the examination room you must give this book to the invigilator. If you do not you may lose all the marks for this paper.

SCOTTISH
QUALIFICATIONS
AUTHORITY

FORMULAE LIST

Circumference of a circle: $C = \pi d$
Area of a circle: $A = \pi r^2$

Theorem of Pythagoras:

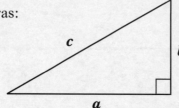

$$a^2 + b^2 = c^2$$

Trigonometric ratios
in a right angled
triangle:

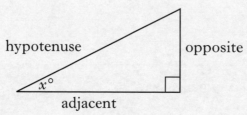

$$\tan x° = \frac{\textbf{opposite}}{\textbf{adjacent}}$$

$$\sin x° = \frac{\textbf{opposite}}{\textbf{hypotenuse}}$$

$$\cos x° = \frac{\textbf{adjacent}}{\textbf{hypotenuse}}$$

Marks

ALL questions should be attempted.

1. The bar graph shows the number of hotels in Southbay awarded grades A to E by the local tourist board.

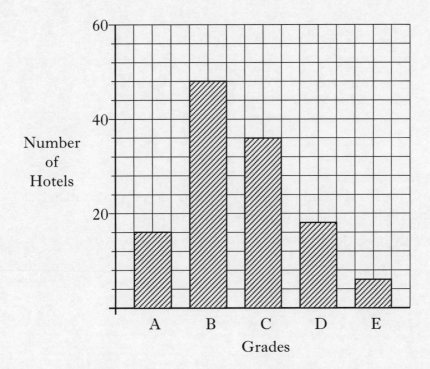

(*a*) How many hotels were awarded an A grade?

1

(*b*) Write down the modal grade.

1

[Turn over

DO NOT WRITE IN THIS MARGIN

Marks

2. The distance from Earth to the Sun is approximately 150 million kilometres. Write this number in standard form. **2**

3. An aeroplane took off from Edinburgh at 0753 and landed in Shetland at 0908. The distance flown by the aeroplane was 295 miles.

Calculate the average speed of the aeroplane in miles per hour. **3**

4. Solve algebraically the equation

$$17y - 12 = 3y + 44.$$

3

Marks

5. A teacher records the number of absences and end of term test mark for each of her students.

The scattergraph shows the results.

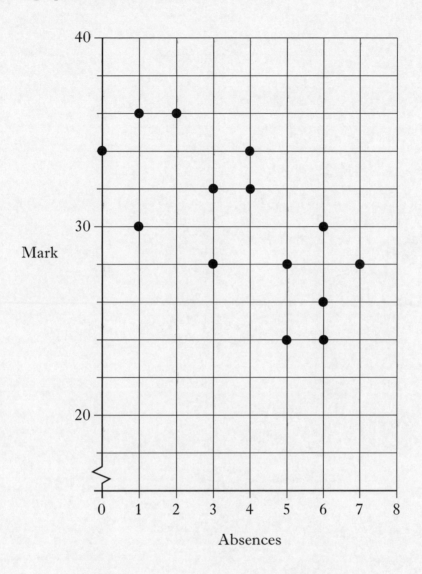

(*a*) Draw a line of best fit through the points on the graph.

1

(*b*) Use your line of best fit to estimate the mark of a student who had 8 absences.

1

[Turn over

Marks

6. (*a*) Multiply out the brackets and simplify

$$3(5p + 3) - 2p.$$

2

(*b*) Factorise $21 - 14m.$

2

Marks

7. The weights of two groups of ten people are to be compared.

Listed below are the weights (in kilograms) of the ten people in group A.

$$64 \quad 71 \quad 73 \quad 66 \quad 69 \quad 78 \quad 77 \quad 75 \quad 76 \quad 71$$

(*a*) Find the median.

2

(*b*) Find the range.

2

(*c*) For the ten people in group B the median is 76 and the range is 20.

Make **two** comments comparing the weights of the people in group A and group B.

2

[Turn over

DO NOT WRITE IN THIS MARGIN

Marks

8. Sam invests £7600 in a bank account.

- The rate of interest is 4·8% per annum.
- The bank deducts 20% tax from the interest.

Calculate the interest Sam receives for one year after tax has been deducted. **3**

Marks

9. Phil is making a wooden bed frame.

The frame is rectangular and measures 195 centimetres by 95 centimetres.

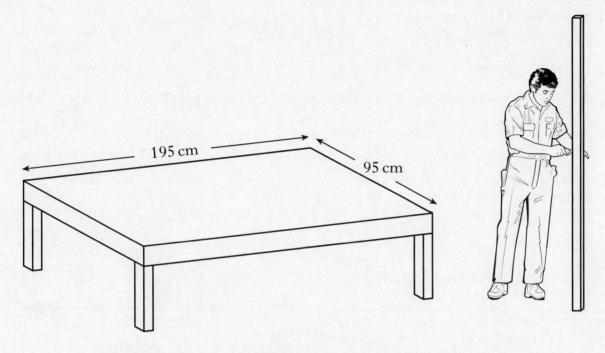

195 cm

95 cm

To make the frame rigid, Phil is going to add a piece of wood along one of its diagonals.

He has a piece of wood 2·2 metres long.

Is this piece of wood long enough to fit along the diagonal?

Give a reason for your answer.

Do not use a scale drawing.

4

[Turn over

Marks

10. Curtis flew from New York to London where he changed 1400 dollars into pounds.

He spent £650 in London and then changed the rest into euros before travelling to Paris.

How many euros did Curtis receive?

Exchange Rates	
$	£1 = 1·75 dollars
€	£1 = 1·38 euros

3

Marks

11. Three roads form a right angled triangle as shown in the diagram.

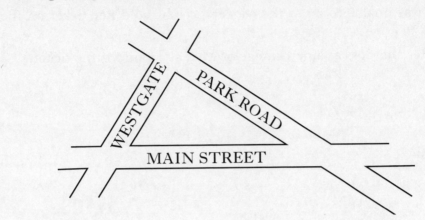

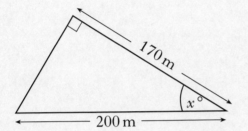

- Main Street is 200 metres long.
- Park Road is 170 metres long.
- The angle between Westgate and Park Road is $90°$.

The size of the angle between Main Street and Park Road is $x°$.

Calculate x.

Give your answer to **one decimal place**.

4

[Turn over

Marks

12. Pamela paid £40 for a concert ticket.

   She was unable to go to the concert, so she sold her ticket on the Internet for £26.

   Express her loss as a percentage of what she paid for the ticket.

4

Marks

13. The diagram below shows a birthday card.

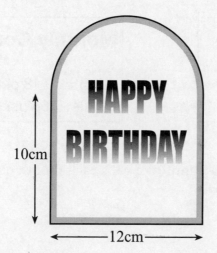

10cm

12cm

The card consists of a rectangle and a semi-circle.

There is gold ribbon all round the border of the card.

Calculate the total length of gold ribbon needed for this card.

Give your answer to the **nearest centimetre**.

5

[Turn over for Question 14 on *Page fourteen*

DO NOT WRITE IN THIS MARGIN

Marks

14. The tariffs shown below are available when buying a mobile phone.

Pay As You Go
Calls: 14p per minute

Monthly Contract
Rental: £18 per month
Calls: 6p per minute

(*a*) Find the cost of using 200 minutes of calls each month on the:

(i) Pay As You Go tariff;

(ii) Monthly Contract tariff. **2**

(*b*) Nick and Amy have mobile phones.

Nick is on Pay As You Go and Amy has a Monthly Contract.

In April:

- the cost to each was exactly the same
- Nick used the same number of minutes as Amy.

How many minutes was this? **3**

[*END OF QUESTION PAPER*]

ADDITIONAL SPACE FOR ANSWERS

DO NOT
WRITE IN
THIS
MARGIN

ADDITIONAL SPACE FOR ANSWERS

[BLANK PAGE]

FOR OFFICIAL USE

X100/101

Total mark

NATIONAL
QUALIFICATIONS
2008

TUESDAY, 20 MAY
1.00 PM – 1.35 PM

MATHEMATICS
INTERMEDIATE 1
Units 1, 2 and 3
Paper 1
(Non-calculator)

Fill in these boxes and read what is printed below.

Full name of centre

Town

Forename(s)

Surname

Date of birth

Day　　Month　　Year　　Scottish candidate number

Number of seat

1　**You may NOT use a calculator.**

2　Write your working and answers in the spaces provided. Additional space is provided at the end of this question-answer book for use if required. If you use this space, write clearly the number of the question involved.

3　Full credit will be given only where the solution contains appropriate working.

4　Before leaving the examination room you must give this book to the invigilator. If you do not you may lose all the marks for this paper.

Use blue or black ink. Pencil may be used for graphs and diagrams only.

FORMULAE LIST

Circumference of a circle: $C = \pi d$

Area of a circle: $A = \pi r^2$

Theorem of Pythagoras:

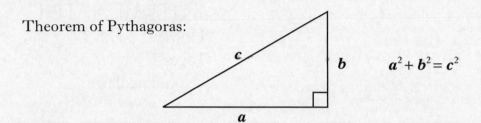

$$a^2 + b^2 = c^2$$

Trigonometric ratios
in a right angled
triangle:

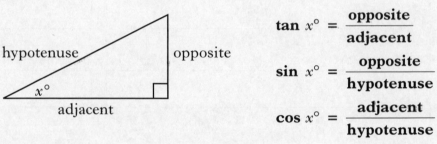

$$\tan x° = \frac{\text{opposite}}{\text{adjacent}}$$

$$\sin x° = \frac{\text{opposite}}{\text{hypotenuse}}$$

$$\cos x° = \frac{\text{adjacent}}{\text{hypotenuse}}$$

Marks

ALL questions should be attempted.

1. (*a*) Find $2 \cdot 685 - 0 \cdot 29$.

1

(*b*) Find 14×3000.

1

(*c*) Find $5 \cdot 45 \div 5$.

1

2. Sandra works night shift. One night she started work at 2235 and finished at 0715 the next morning.

How long did Sandra's shift last?

1

[Turn over

Page three

Marks

3. The diameter of a red blood cell is $6{\cdot}5 \times 10^{-3}$ millimetres.

Write this number in full.

2

4. A plumber charges £20 for being called out to a job, plus £12 **for each 15 minutes** he takes to do the job.

How much does he charge for a job which takes 2 hours?

2

Marks

5. A building company employs 70 staff.

   The number of staff absences during the last year is shown in the frequency table below.

Number of Absences (Days)	Frequency
0	7
1	21
2	18
3	11
4	8
5	5
Total	70

   (a) Find the probability of choosing a member of staff who had no absences.

   1

   (b) Complete the table below **and** calculate the mean number of absences.

Number of Absences (Days)	Frequency	Number of Absences × Frequency
0	7	0
1	21	21
2	18	36
3	11	
4	8	
5	5	
Total	70	

   3

6. Frances is on holiday. She wants to book some of the excursions shown in the advert below.

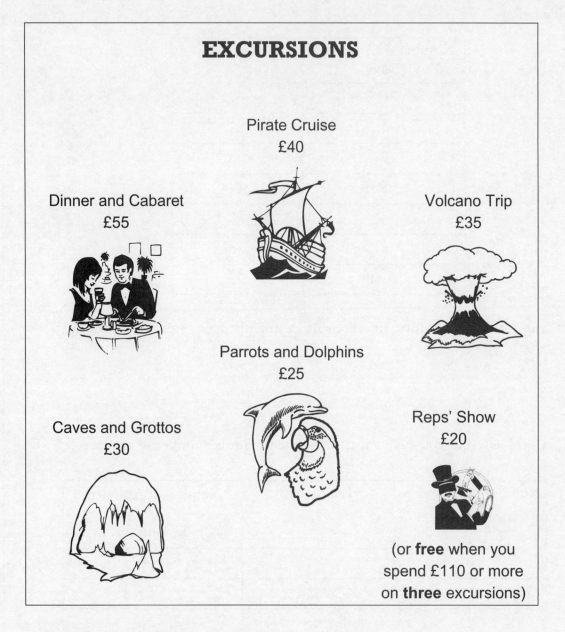

EXCURSIONS

Pirate Cruise
£40

Dinner and Cabaret
£55

Volcano Trip
£35

Parrots and Dolphins
£25

Caves and Grottos
£30

Reps' Show
£20

(or **free** when you
spend £110 or more
on **three** excursions)

- Frances wants to book **four** different excursions.
- She can afford to spend a **maximum of £120**.
- She gets a **free** ticket for the Reps' Show when she spends £110 or more on **three** excursions.

Marks

6. (continued)

Two combinations of **four** excursions that Frances can afford are shown in the table below.

Dinner and Cabaret £55	55							
Pirate Cruise £40		40						
Volcano Trip £35		35						
Caves and Grottos £30	30							
Parrots and Dolphins £25	25	25						
Reps' Show £20 or Free	Free	20						
Total Price	£110	£120						

Complete the table to show **all** possible combinations that Frances can afford.

3

7. Solve algebraically the equation

$$7m - 8 = 40 + m.$$

3

[Turn over

Marks

8. (*a*) Complete the table below for $y = 2 \cdot 5x - 3$.

x	−2	0	2	4
y			2	

2

(*b*) Draw these **two** lines on the grid:

(i) $y = 2 \cdot 5x - 3$;

(ii) $y = 3$.

3

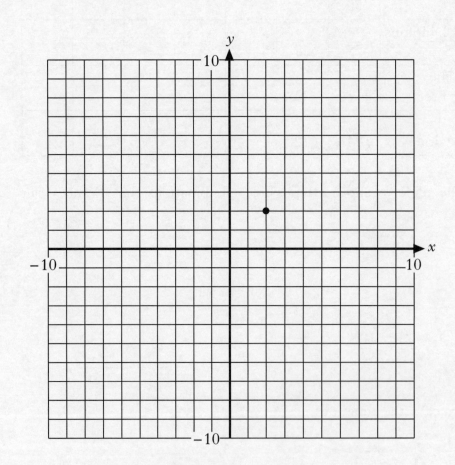

DO NOT WRITE IN THIS MARGIN

Marks

9. Evaluate $x^2 - y$ when $x = -8$ and $y = 73$. **3**

10. Jamie invests £1440 in a savings account.

The rate of interest is 5% per annum.

Calculate the interest he should receive after 3 months. **4**

[*END OF QUESTION PAPER*]

DO NOT
WRITE IN
THIS
MARGIN

ADDITIONAL SPACE FOR ANSWERS

FOR OFFICIAL USE

Total mark

X100/103

NATIONAL
QUALIFICATIONS
2008

TUESDAY, 20 MAY
1.55 PM – 2.50 PM

MATHEMATICS
INTERMEDIATE 1
Units 1, 2 and 3
Paper 2

Fill in these boxes and read what is printed below.

Full name of centre

Town

Forename(s)

Surname

Date of birth

Day   Month   Year

Scottish candidate number

Number of seat

1 **You may use a calculator.**

2 Write your working and answers in the spaces provided. Additional space is provided at the end of this question-answer book for use if required. If you use this space, write clearly the number of the question involved.

3 Full credit will be given only where the solution contains appropriate working.

4 Before leaving the examination room you must give this book to the invigilator. If you do not you may lose all the marks for this paper.

Use blue or black ink. Pencil may be used for graphs and diagrams only.

FORMULAE LIST

Circumference of a circle: $C = \pi d$
Area of a circle: $A = \pi r^2$

Theorem of Pythagoras:

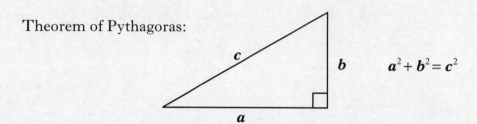

$a^2 + b^2 = c^2$

Trigonometric ratios
in a right angled
triangle:

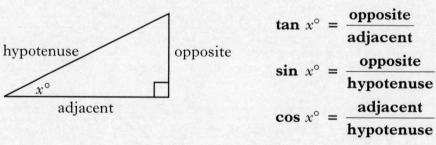

$$\tan x^\circ = \frac{\text{opposite}}{\text{adjacent}}$$

$$\sin x^\circ = \frac{\text{opposite}}{\text{hypotenuse}}$$

$$\cos x^\circ = \frac{\text{adjacent}}{\text{hypotenuse}}$$

Marks

ALL questions should be attempted.

1. (*a*) On the grid below plot the points A(–2,4), B(–4,–1) and C(1,–3).

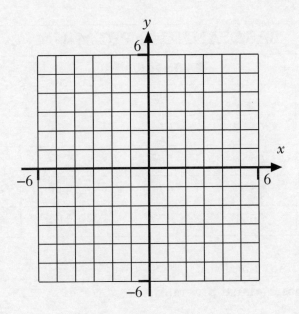

2

(*b*) Plot the point D so that shape ABCD is a square.

1

[Turn over

Page three

Marks

2. The table below shows the basic annual premiums charged for car insurance by an insurance company.

   The basic premium depends on the area where the driver lives and the group their car belongs to.

AREA	BASIC ANNUAL PREMIUM				
	CAR GROUP				
	1	2	3	4	5
A	£428	£517	£613	£725	£838
B	£497	£555	£659	£779	£898
C	£525	£598	£712	£841	£975
D	£540	£651	£775	£915	£1055

(a) Lynn's car is in group 4 and she lives in area C.

   Write down her basic annual premium.

1

Drivers who do not make a claim on their insurance receive a discount on their basic annual premium as shown in the table below.

Number of years without a claim	1	2	3	4 or more
Discount	30%	40%	55%	67%

(b) Lynn has not made a claim for 4 years.

   How much will it cost her to insure her car?

2

DO NOT WRITE IN THIS MARGIN

Marks

3. (*a*) Multiply out the brackets and simplify

$$4(5u - 2) + 15.$$

2

(*b*) Factorise $9c + 24$.

2

[Turn over

Marks

4. A grass lawn is treated with weedkiller.

The lawn is split into twenty squares each of the same area.

Ten of the squares are treated with Weedclear.

Three weeks later the number of weeds in each of these squares is as follows:

$$3, \quad 4, \quad 6, \quad 2, \quad 1, \quad 7, \quad 2, \quad 1, \quad 1, \quad 3.$$

(*a*) Find the median.

2

(*b*) Find the range.

1

The other ten squares are treated with Noweed.

For these squares the median is 2 and the range is 10.

(*c*) Make **two** comments comparing the number of weeds in squares treated with Weedclear and Noweed.

2

Marks

5. Ross drove 190 miles from Preston to Edinburgh in 3 hours 30 minutes.

During the first part of his journey he drove for 2 hours at an average speed of 68 miles per hour.

Find the average speed in miles per hour for the rest of his journey.

4

[Turn over

Marks

6. Some biology students were doing a project on "creepy crawlies". The pie chart shows the different types of creepy crawlies that the students collected from a garden.

CREEPY CRAWLIES

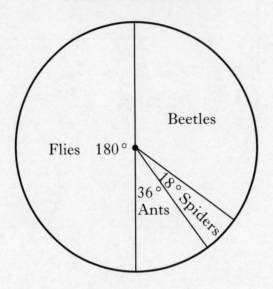

The students collected 220 creepy crawlies altogether.

How many of them were beetles?

3

Marks

7. A farmer is building a wire fence around a field.

The fence has heavy posts at the corners.

Each corner post is supported by a stake as shown in the diagram.

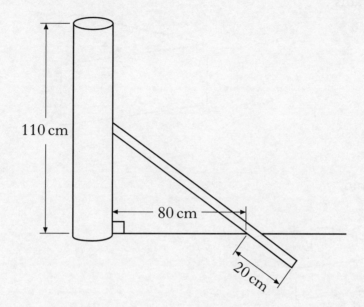

- The corner post is 110 centimetres high.
- The stake meets the corner post halfway up.
- The stake meets the ground 80 centimetres from the foot of the corner post.
- 20 centimetres of the stake is below ground level.

Calculate the length of the stake.

Do not use a scale drawing.

4

[Turn over

Marks

8. Shown below are two pieces of cheese.

The weight of each piece is proportional to its volume.

Piece A has a volume of 400 cubic centimetres.
It weighs 480 grams.

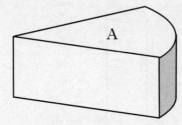

Piece B is a cuboid.

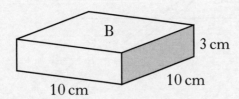

Find the weight of piece B.

4

Marks

9. The table shows the ticket prices for a theme park in France.

   The prices are given in euros.

Ticket	Adult price	Child price
Bronze (valid 1 day)	€50	€40
Silver (valid 2 days)	€90	€75
Gold (valid 3 days)	€110	€85

   Gavin buys silver tickets for two adults and one child.

   Find the total cost, in pounds and pence, of buying these tickets if the exchange rate is £1 = 1·39 euros.

3

[Turn over

Marks

10. Solve algebraically the inequality

$$\tfrac{1}{2}y + 3 > 13.$$

2

11. Calculate the area of the rectangle shown below.
Do not use a scale drawing.

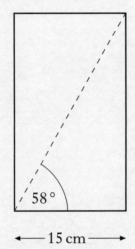

58°

←— 15 cm —→

4

Marks

12. Use the formula below to find the value of T when $r = 2\cdot6$ and $s = 1\cdot4$.

$$T = \frac{rs}{r+s}$$

3

[Turn over

Marks

13. Sergei has been training to run a marathon.

Since he started training his weight has dropped from 80 kilograms to 74 kilograms.

Express his weight loss as a percentage of his original weight.

4

Marks

14. The diagram below shows part of a garden which is being watered from a sprinkler.

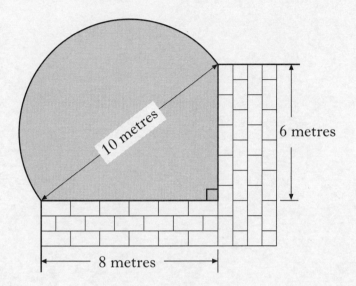

10 metres

6 metres

8 metres

The area being watered is in the shape of a semi-circle and a right angled triangle.

Calculate the area being watered.

4

[END OF QUESTION PAPER]

DO NOT
WRITE IN
THIS
MARGIN

ADDITIONAL SPACE FOR ANSWERS

Pocket answer section for
SQA Mathematics Intermediate 1: Units 1, 2 and 3
2004–2008

© 2008 Scottish Qualifications Authority/Leckie & Leckie, All Rights Reserved
Published by Leckie & Leckie Ltd, 3rd Floor, 4 Queen Street, Edinburgh EH2 1JE
tel: 0131 220 6831, fax: 0131 225 9987, enquiries@leckieandleckie.co.uk, www.leckieandleckie.co.uk

Mathematics Intermediate 1
Units 1, 2 and 3
Paper 1 (Non-calculator)
2004

1. (a) £69

    (b) 60

    (c) 200

2. £599

3. 12m³

4. (a) −1·5

    (b) 8

    (c) Invergow colder than Abergrange. Temperatures vary more in Invergow.

5. $x = 6$

6.

Carnation	Daffodil	Lily	Iris	Rose	Total Price
		✓	✓	✓	£11·50
✓	✓			✓	£10·00
✓		✓		✓	£10·50
	✓	✓	✓		£10·50
	✓		✓	✓	£11·00
	✓	✓		✓	£12·00

7. (a) 3·9

    (b) 5

8. (a)

x	-2	2	7
y	5	1	-4

8. (b)

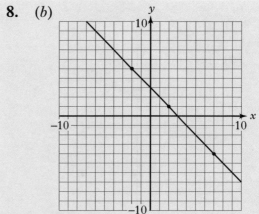

9. 20

10. 15

Mathematics Intermediate 1
Units 1, 2 and 3
Paper 2
2004

1. $\frac{10}{2000}$

2. (a)

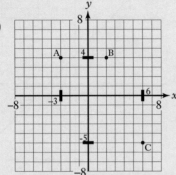

  (b)

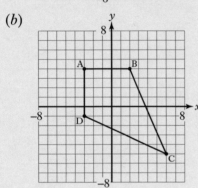

      D(−3,−1)

3. 48 mph

4. $n < 5$

5. (a)

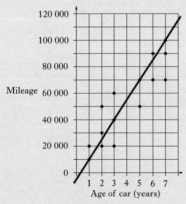

  (b) 55 000 miles

6. (a) $8 + 3t$

  (b) $5(2y - 7)$

7. £36 000

8. (a) £60

  (b) 28·3 inches

9. One of:
- **Yes** - it costs €39·50 in Scotland
- **Yes** - it costs £25·33 (or £25·32) in Spain

10. 53°

11. (a) 3·25%

  (b) It went down.

  (c) £8·75

12. 9·3…

13. 45%

14. 1·46 m²

Mathematics Intermediate 1
Units 1, 2 and 3
Paper 1 (Non-calculator)
2005

---

1. (*a*) 3·87

   (*b*) £900

2. 8.50 am

3. 120

4. 5·67

5. $a = 9$

6.

Digital Camera £95	Scanner £75	Printer £70	Cordless Keyboard £45	Pair of Speakers £40	Total Value
✓	✓				170
✓		✓			165
✓			✓	✓	180
	✓	✓	✓		190
	✓		✓	✓	160

7. (*a*) 9, 5, -3

   (*b*)

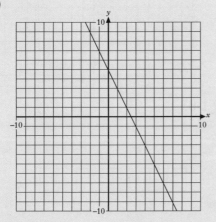

8. (*a*) $81

   (*b*) £1 = $1·60

9. (*a*) 0·007

   (*b*) $7·1 \times 10^{-4}, \dfrac{7}{1000}, 0·069$

10.(*a*) −3

   (*b*)

1	−6	−1
−4	−2	0
−3	2	−5

Mathematics Intermediate 1
Units 1, 2 and 3
Paper 2
2005

1. 166,000 cm³

2. £2410

3. (a) 3h 15m

    (b) 40 mph

4. $t > 8$

5. (a) 41 years

    (b) 22 years

    (c) eg Kestrels are older

6. (a) $3n + 28$

    (b) $3(5 + 2x)$

7. (a) 75

    (b) 72·5

    (c) $\dfrac{3}{12}$

8. 70%

9. 490

10. 18·6 cm

11. 419·8 m or 419·9 m

12. 180

13. 8·5 cm

14. £52

15. (a) 9

    (b) 73

Mathematics Intermediate 1
Units 1, 2 and 3
Paper 1 (Non-calculator)
2006

1. 3·62

2. 167

3. 4 m 10 s

4. £162

5. £46

6. $n = 6$

7. (a)

x	−3	0	2
y	−7	2	8

    (b)

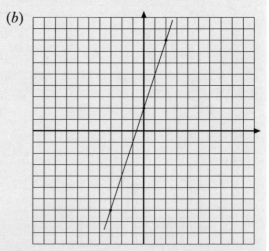

8. £1·05

9. 6

10. (a)

4	−6	−2	−8

    (b)

−6	5	−1	4

    (c)

1	−4	−3	−7

Mathematics Intermediate 1
Units 1, 2 and 3
Paper 2
2006

1. 12430 pesos

2. $5\cdot4 \times 10^{-6}$

3. $t > 9$

4. 455

5. (a) 15

   (b) $\dfrac{3}{40}$

   (c) $16\cdot3$

6. 240 litres

7. (a) $2x - 5y$

   (b) $4(2d + 3)$

8. (a) $10\cdot5$

   (b) 15

   (c) More cars on Monday.
        Number of cars vary more on
        Monday.

9. $1\cdot3$ m

10.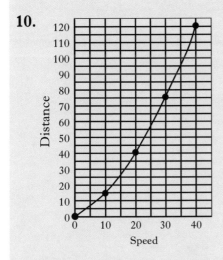

11. £78

12. 25 m

13. 20%

14. $21\cdot9\,\text{m}^2$

Mathematics Intermediate 1
Units 1, 2, and 3
Paper 1 (Non-calculator)
2007

1. (a)   19·22

    (b)   0·47

    (c) $\dfrac{57}{1000}$

    (d)   £288

2. £61·20

3. $a < 9$

4.
$$\begin{array}{r} 108 \\ 60 \\ \hline \text{Total} = 390 \end{array}$$

    7·8 minutes

5. (a) $-7, -3, 9$

    (b)

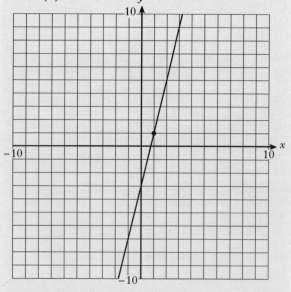

6. 8 cm

7. (a) -8

    (b)   17

8.

	Lamp 100 watts	Computer 200 watts	Games Machine 400 watts	Microwave 700 watts	Heater 1000 watts	Kettle 2300 watts	Total Watts
	✓	✓	✓		✓		1700
				✓		✓	3000
	✓	✓		✓	✓		2000
	✓		✓	✓	✓		2200
		✓	✓	✓	✓		2300
	✓	✓	✓	✓			1400

9. 54

10. bag 1: $\frac{3}{5} = \frac{9}{15}$

     bag 2: $\frac{8}{15}$

     so bag 1

Mathematics Intermediate 1
Units 1, 2, and 3
Paper 2
2007

1.   (a)   16

    (b)   B

2. $1 \cdot 5 \times 10^8$

3.   236 mph

4. $y = 4$

5.   (a)

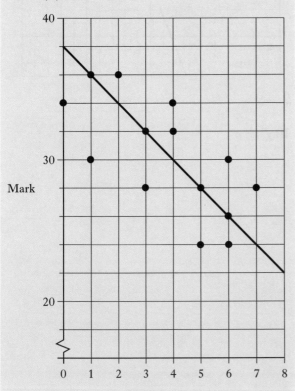

Absences

    (b)   Your answer must be consistent with your line of best fit. The correct answer for the given line is 22.

6.   (a) $13p + 9$

    (b) $7(3 - 2m)$

7.   (a)   72 kg

    (b)   14 kg

    (c)   Group B heavier and weights vary more

8.   £291·84

9.   Yes, since 217 cm (length of diagonal) < 220 cm (length of wood)

10.   €207

11.   31·8°

12.   35%

13.   51 cm

14.   (a)   (i) £28
           (ii) £30

    (b)   225 minutes (cost = £31·50)

Mathematics Intermediate 1
Units 1, 2 and 3
Paper 1
2008

1. (*a*) 2·395

   (*b*) 42 000

   (*c*) 1·09

2. 8 hours 40 minutes

3. 0·0065

4. £116

5. (*a*) $\frac{7}{70}$

   (*b*) 2·1

6.

Dinner and Cabaret £55	55	55	55		
Pirate Cruise £40	40			40	
Volcano Trip £35		35	35		35
Caves and Grottos £30		30		30	30
Parrots and Dolphins £25	25		25	25	25
Reps' Show £20 or Free	Free	Free	Free	20	20
Total Price	120	120	115	115	110

7. m = 8

8. (*a*)

x	−2	0	2	4
y	−8	−3	2	7

   (*b*)

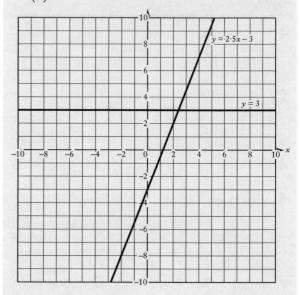

9. −9

10. £18

Mathematics Intermediate 1
Units 1, 2 and 3
Paper 2
2008

1.

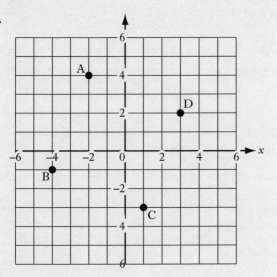

2. (a) £841

   (b) £277·53

3. (a) $20u + 7$

   (b) $3(3c + 8)$

4. (a) 2·5

   (b) 6

   (c) Less weeds remain with Noweed. Number of remaining weeds varies more with Noweed.

5. 36 mph

6. 77

7. 117 cm

8. 360 grams

9. £183·45

10. $y > 20$

11. 360 cm²

12. 0·91

13. 7·5%

14. 63 m²

Official SQA answers to 978-1-84372-651-7
2004–2008